Eric Aley

Metaphorical Visualization

Eric Aley

Metaphorical Visualization

Using Multidimentional Metaphors to Visualize Data

VDM Verlag Dr. Müller

Copyright © 2007 VDM Verlag Dr. Müller e. K. and licensors
All rights reserved. Saarbrücken 2007
Contact: info@vdm-verlag.de
Cover image: www.purestockx.com
Publisher: VDM Verlag Dr. Müller e. K., Dudweiler Landstr. 125 a, 66123 Saarbrücken, Germany
Produced by: Lightning Source Inc., La Vergne, Tennessee/USA
          Lightning Source UK Ltd., Milton Keynes, UK

Copyright © 2007 VDM Verlag Dr. Müller e. K. und Lizenzgeber
Alle Rechte vorbehalten. Saarbrücken 2007
Kontakt: info@vdm-verlag.de
Coverbild: www.purestockx.com
Verlag: VDM Verlag Dr. Müller e. K., Dudweiler Landstr. 125 a, 66123 Saarbrücken, Deutschland
Herstellung: Lightning Source Inc., La Vergne, Tennessee/USA
          Lightning Source UK Ltd., Milton Keynes, UK

ISBN: 978-3-8364-2878-1

ABSTRACT

Real-Time Metaphorical Visualization of

Multi-Dimensional Environmental Data. (May 2006)

Eric Brian Aley, B.E.D., Texas A&M University

Chair of Advisory Committee: Prof. Carol LaFayette

This research explores the process of reformulating multiple data sets into metaphorical representations. The representations must coherently intertwine into a multi-level metaphor that constrains their forms. A working installation has been created, using the natural environment as a metaphor for the built environment. Numerical measurements of weather conditions inside of Texas A&M's Langford architectural building are translated into visual metaphors that map to the metaphor of a landscape – rainfall, wind strength, grass color, and lightning – to visually describe the state of the building in real time.

ACKNOWLEDGEMENTS

This work wouldn't have been possible without the help and support of quite a few people.

My committee members, Andruid Kerne and Louis Tassinary, both gave me their experience and lent me a great deal of equipment, as well as both setting me on the path towards this project from opposite directions – Andruid in his unusual and innovative courses, Lou by finding and getting me access to Langford's electrical records just in case I could do something with them.

As my committee chair, Carol LaFayette put up with good grace with my indecisive and sometimes frustrating methods of developing the project. She also lent equipment and software development tips that helped refine the final work. As a teacher, she led the entire VIZA 644 class in finishing the visual aspects of the project, and I owe virtually all of the compositional improvements to her and all of those students.

Special thanks must be given to Lauren Simpson, who left her laptop with me for the entire book-writing semester, and to my roommates Christina Croxell and Alethea Bair who tolerated my irregular working patterns.

Finally, none of this could have been done without the support of my parents, who let me stay in school long enough to finish two degrees.

TABLE OF CONTENTS

LIST OF FIGURES

CHAPTER I

INTRODUCTION

This work presents an approach to the creation of multi-dimensional metaphorical data visualization, as well as a functional demonstration. In order to understand the difficulties of multi-dimensional visualization, the different types of two-, three-, or more-dimensional graphs should be discussed. Those are covered here, as well as an introduction of the environmental measurements that are used in the final work.

The final demonstration of this approach is a multi-dimensional visualization of environmental data taken from the Langford architectural complex at Texas A&M University, where the values of four variables, namely temperature, humidity, airflow through the ventilation system, and amount of electricity use, will be sampled and visualized.

1.   Visualization

"Excellence in statistical graphics consists of complex ideas communicated with clarity, precision, and efficiency." [1]

A graph can transform vast strings of numbers into a single image that quickly shows trends and anomalous values hidden in the data. Well-designed graphs can provide compelling evidence of correlations or differences between groups of numbers. A table with a dozen dimensions of data can be succinctly compressed into one picture

---

This work follows the style and format of *IEEE Transactions on Visualization and Computer Graphics*.

with almost no numbers but very high readability.

Reading an array of numbers on their own, much less several of arrays at once, quickly becomes a daunting task. The numbers lose relevance, trends get lost in the noise, and the usefulness of a quick glance vanishes.

Additionally, numerical statistical calculations can be misleading. Anscombe's quartet is a cautionary example of four different data sets of X and Y values whose linear models are exactly the same. The number of samples, the means, the equation of the regression line, the sum of squares, the regression sum of squares, correlation coefficient, and more are all identical across the four data sets. Only when the sets are visualized do their enormous differences appear [1].

1.1. 2D Graphs

Graphical displays have always been used to show relationships between numbers. Bar, pie, and Cartesian graphs are staples of numerical analysis, relating magnitudes of numbers with the size of their bar, or slice, or height. Depth maps of the ocean correlate the deepness of the blue in the image with the deepness of the water at that location. Each of these metaphors is simple – so simple that at first they hardly seem like metaphors at all. The meaning of a darker strip of blue in a drawing of a body of water is so intuitively understandable that the map's color key is hardly needed for anything besides the actual quantitative information [2].

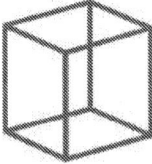

Figure 1. The Necker Cube, a demonstration of the problems projecting data from higher dimensions to displays of low dimensions. The amount of information present in this 2D drawing of a cube is insufficient to determine 3D spatial information, such as which edge is in front.

1.2. Multi-Dimensional Graphs

Generally speaking, a graph will end up in two dimensions: printed on paper; displayed on a computer screen; or projected onto a wall. This is straightforward with two-dimensional graphs, but introduces problems with three or more dimensions, because information is lost along with the extra dimensions (Figure 1). With 3D projection the missing dimension can often be replaced with a different third dimension – time – and an animation of the visualization rotating or otherwise moving will supply the extra information. Multidimensional graphing usually is limited to the visualization of more than three data dimensions, but a few three-dimensional methods are included here because animation is not always feasible.

Multidimensional visualization comes in four basic forms, which will be summed in brief here, and described in more depth in following paragraphs. Color is a straightforward addition to nearly any visualization method, allowing color to represent one or more additional dimensions. The extra dimensions can be drawn over each other such that all the data they represent now lie in one plane, as is done with parallel graphs.

The large dimension set can be projected into two or three dimensions through careful algorithms meant to allow the multidimensional structures to appear, as with Andrews curves. And finally the data points in many 2D graphing methods can be drawn as "glyphs," icons that change different features within them depending on the extra dimensions of the data they represent, as with nested and star graphs, and Chernoff faces [3].

Projection techniques merely display the data as two-dimensional, without regard for the loss of information. A variation on this simplest form of projection involves rotating the data multidimensionally while only displaying the flattened 2D projection, which can reveal unusual structures through time. Andrews curves are a more complex technique, in which each data point is put through an equation that creates a unique 2D curve for each set of data. The algorithm is useful because multidimensional clusters will tend to have similar curves, however the abstraction renders them useless to the layman.

Parallel graphs are created when the extra dimensions are drawn on top of each other, making the orthogonal axes parallel (Figure 2). If all the extra dimensions are made parallel by stacking them on top of each other, data clusters can be seen quickly (Figure 3). However, overlapping data lines also can obscure each other, making it impossible to trace a single data dimension. Colors or line styles are often used to help distinguish dimensions.

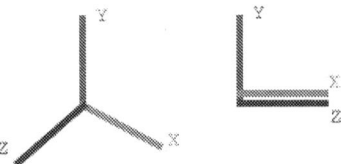

Figure 2. A typical icon of 3D axes implies three dimensions through its angles, and the assumption that they're actually orthogonal to each other. A parallel graph would rotate all the dimensions beyond the first two to be parallel to each other. This points the X and Z axes both in the same direction so they can be displayed in two dimensions.

Glyph plotting involves replacing data points with icons that represent the remaining dimensions in some way. Conceptually, the simplest type of glyph plotting is nested graphing, where a data point in the first level of display serves as the origin of a second coordinate system, incorporating the next set of dimensions to describe the next group of information. Though literal and effective, nested graphs are visually confusing and lack expandability.

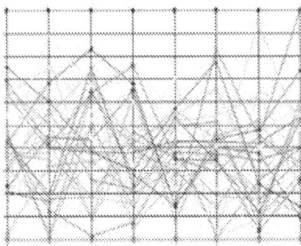

Figure 3. A parallel graph using color to attempt to distinguish between the dimensions [3].

Star graphs are a more compact form of glyph plotting, where the data points are drawn as symbols iconic of data dimensions [4]. They plot data along two axes, as with parallel graphs and two-dimensional graphing, but each data point is drawn as the intersections of additional axes to bring in extra dimensions of data. If the minor axes are connected, the data points become star-shaped, unique to their combination of data, hence the name. While star graphs present information in a very compact form, they lack contextual connections between the data and the visualization. Without a great deal of practice, reading them is impossible due to the lack of reference between the star shapes and the actual data. Trends are also difficult to follow because of the individuality of the shapes.

Chernoff faces are similar to star graphs, except that rather than using literal data axes the data points are encoded into metaphors of faces [5]. Additional dimensions of data can be described by, for example, the shape of the face, length of the nose, or size of the eyes. The use of faces as a vehicle for data representation arose from the recognition that humans are particularly talented at recognizing differences between faces. By arbitrarily connecting data sources to aspects of the face, it is hoped that trends within the graph can be quickly read. Like star graphs, Chernoff faces provide no context to the mappings of their dimensions. Their use requires referencing a key in order to connect each feature of the face with their data dimension.

The field of medicine is beginning to solve some visualization problems with multidimensional techniques. Though limited by a need to be more literal than nested coordinates or Chernoff faces, there nevertheless have been many improvements over the traditional series of two-dimensional images describing individual slices of the person being scanned. The simplest of these merely stacks the image slices and interpolates partially transparent 3D models that can be rotated and color-coded for the doctor's convenience. More complex are techniques like the acoustic modeling being used to improve the design of hearing aids by mapping sound pressure levels across a human head to show how audible direction-finding is processed [6].

1.3. Real-Time Visualizing

The traditional emphasis in the multidimensional visualization field is on interactive responsiveness and changing levels of detail. Users can dynamically sift through data and effectively shrink the dimensions or range to a more manageable size.

Although all visualizations are inherently somewhat metaphorical, it is useful to distinguish between more statistical techniques, such as bar or star graphs, and the more abstract Chernoff faces. Statistical techniques translate directly into numbers, as in the case of parallel graphs, where each data point corresponds directly with the quantified information. On the other hand, while metaphorical visualizations like Chernoff faces generally limit the numerical precision displayed, their goal is fast interpretation and intuitive understanding of the relationships between the different samples.

1.4. Ambient Visualizations

Ambient displays are usually real-time one-dimensional mappings between something that must be constantly monitored to a display that is always present but never distracting [7].

Sound has one major advantage over sight when ambient devices are designed, while sight has several over sound. Audio cues do not require a physical orientation of the viewer; they can be heard and responded to while facing in any direction, while visual indicators must be within the user's cone of vision. For one-dimensional tasks, particularly tasks in the style of warning sirens where audio cues have been used for decades, audio is probably more effective than visual indicators. Once one gets beyond a few dimensions, however, vision becomes most appropriate, as "the sense with by far the largest bandwidth." [8]

As more dimensions are added, visual cues quickly show their strength. Sound is heard all at once; as more dimensions are slipped into the signal, it becomes increasingly difficult to isolate different aspects of it – the signal can start to sound like noise. Sight is localized: more information can be added in different places, with the eyes capable of isolating different pieces as needed.

Ambient displays are particularly suited for metaphorical techniques; the preference is always on intuitive comprehension and a quick understanding of the meaning of the data rather than the specifics. This meshes perfectly with a metaphorical display's strengths, while rendering the low numerical precision irrelevant.

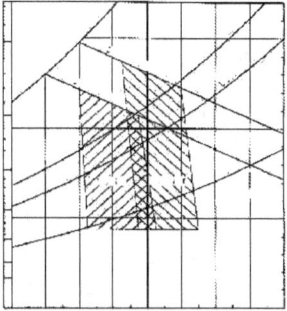

Figure 4. Graph demonstrating winter and summer comfort zones at the intersections of humidity, dry- and wet-bulb temperatures.

2.   Environmental Measurements

One of the primary purposes of the built environment is to create comfortable living and working environments for people. This is accomplished via the simple (a roof to stave off rain and sun, or insulation to hold back the heat or chill) and the complex (automatic air conditioners or adaptive ventilation systems). But what exactly is comfortable?

Temperature and humidity have an established psychrometric range where most people find the combination of the two comfortable (Figure 4). The comfort zone shifts between seasons as people acclimate to the weather, but generally spans from 20 to 26 degrees Celsius, and 30 to 80 percent relative humidity [9].

VAV ventilation systems, or variable air volume systems, alter the amount of airflow depending on specific conditions or operator requests in order to maximize occupant comfort. Generally speaking, the systems are keyed to the air temperature outside the building. Between about 20 and 65 degrees, air movement remains constant.

Above and below those threshold, and there is a linear increase in the volume of air being pushed through the system; the hotter or colder it is outside, the faster the air moves. Because of this, measuring the airflow through a functional VAV system is as easy as checking the temperature outside the building [10].

Finally, the amount of electricity consumed by a building will fluctuate directly proportional to the amount of use it is experiencing. The primary electrical loads of a building are heating and cooling, lighting, and in the case of office or academic buildings, computers; all of which are directly correlated with the number of occupants.

These environmental measurements will all be done inside the Langford architectural complex. Langford is equipped with a VAV system calibrated to change according to the outside temperature, as well as an extensive network of sensors monitoring, among other things, the amount of electricity consumption.

CHAPTER II

HISTORY

"Metaphor is pervasive in everyday life, not just in language but in thought and action." [11] Our language is filled with figures of speech, where an argument can be described as a battle and health has a peak. It allows for symbols, from subtle details in a great work of art to a world where green means go. Metaphor can be a powerful tool to describe characteristics or pass along information.

Metaphors can be layered into multiple levels, where a base metaphor describes a world that sub-metaphors must fit into. With this kind of setup additional ideas can be added and relationships quickly understood, so long as the base metaphor is adhered to.

A version of the multi-level metaphor idea is often implemented in software, in a system used to generate data, rather than display it. Software designers create a direct link between features of its user interface and the "real world" counterparts, in an attempt to help the program make sense to a layman. This is most evident in the "desktop" metaphor used by nearly all graphical user interfaces.

The desktop metaphor was developed with the intent of simplifying human-computer interactions, beginning with Xerox's Alto and Star systems, the precursors of virtually every visually-based operating system since. During the development of these systems, the idea of creating the user's conceptual model of the software first, before the software was written, was put into use [12]. In this model, vacant screen space is referred to as a desktop, complete with documents, folders, and in- and out-boxes for mail.

Individual programs within the computer still create their own metaphors. Video software designers map the image of a razor, used in a traditional film-editing studio, to the button that splits the video file. The film studio metaphor falls apart when tasks are assigned which computers can do but which can't be done in the physical realm, generally special effects or mathematics-based image manipulation procedures. Frequently this results in the interface stepping away from the metaphor and simply using lists of options inside menus.

Some software gets lost in inappropriate metaphors. Susteen's DataPilot™ program, software to connect a cellular phone with a computer, has an interface shaped like a generic flip model cell phone. As the program starts, the phone appears and flips open; the exit button is placed and colored the way many power/end call buttons are on actual phones, and instead of a number keypad, each button does some process in the software. In this case, the metaphor doesn't make much sense: the program isn't supposed to act like a cell phone, merely talk to one. Opposite the "end" button is the "talk" button, which actually only opens the help screen. The result is nonsensical, ambiguous, and difficult to use, an outcome that metaphorical systems can avoid when designed not to be cute but to work. This illustrates the need for interface metaphors to be appropriate, because otherwise they only serve to distract.

1.   Real-Time Visualization

Stephen Eick wrote on techniques to use data sliders as both interface input devices and informational cues, through multiple schemes that encode data into the background of the slider bar itself [13]. Through color, tick marks, and different interpretations of his data, the average number of frost-free days in the 48 continental states, he provides four examples of ways to utilize the data: as a continuous spectrum of color, as continuous interpolated curves, as a discrete state-by-state scale, and as a density plot, all by simply toggling between the available backgrounds for the input slider. Wells and Tassinary expanded on this, demonstrating the effectiveness not just of displaying information within sliders, but also the added improvements of dynamically changing the displayed information while the sliders are in use [14].

Alex Galloway created CarnivorePE™ as a software application that monitors network traffic in real time and allows artists to build virtual installations with the data it collects. Inspired by the FBI monitoring program "Carnivore," CarnivorePE™ (Personal Edition) explicitly tracks six dimensions: time; the IP address of the sender; the port number of the sender; the IP address of the receiver; the port address of the receiver; and the packet data being sent. The third-party clients then create visualizations of these six dimensions through whatever means the artists desire. "Fuel," by Scott Snibbe, creates a constellation of stars to represent every IP address contacted through the network, where the 2D location of the star, its color, and visual effects applied to it are determined by the IP addresses, port information, and times, while its magnitude is affected by the amount of traffic sent – an implicit seventh dimension that CarnivorePE™ monitors. Thus, every computer will create a unique constellation based on the most-used network resources [15].

2.   Ambient Displays

In 1995 Natalie Jeremijenko created an artistic installation titled Live Wire, in which a hanging string was twitched in time with data traffic across any network it's plugged into, demonstrating how heavy the load is at any moment [16]. This created both a physical demonstration of the network status for network administrators and a visual aid for their irate customers.

Students in MIT's Media Lab expanded on Jeremijenko's concept, building a network monitor tool that mapped their computer network's status to the sound of wind. When conditions were good, an audio clip of a calm breeze filled their room. When a catastrophe was imminent and immediate action needed, it escalated to a full-blown thunderstorm. "This information about their computer network is always available, but never demands direct attention unless there is a problem." [17]

Advised by Hiroshi Ishii, coauthor of "Ambient Displays" [7] and founder of the Tangible Media Group within MIT's Media Lab, AmbientDevices markets several products designed as ambient information sources. The most exemplary of these is the Ambient Orb™: an interfaceless translucent ball that glows in a color determined by information it gathers from a wireless network. In general use the Orb™ cycles from green to red depending on how well or poorly the stock market is doing, but the parameters can be changed via the same wireless network.

A similar company, Violet, has created a competitor to the Ambient Orb™ with many more dimensions available, named Nabaztag™. With four glowing spots and moving ears that can be linked to other Nabaztags for nonverbal communication, this device has a lot more simultaneous display capabilities, but also loses the elegance of the Ambient Orb. The ears are perhaps the most interesting addition: they can be linked to another Nabaztag™ around the world such that when a user manipulates the ears of their own the ears of the others move to match, creating a remote sense of presence.

Both Hiroshi Ishii's "AmbientROOM" [7] and Hideaki Ogawa's "air" [18] implement the same type of ambient presence notification, whereby two remote locations are connected through a network. By causing a change in the state of one of the installations, the participant at that location triggers a similar effect at the other site, transmitting a nonverbal message that implies both their whereabouts and that the other person is on their mind.

CHAPTER III

METHODOLOGY

This chapter describes the process used to create a system of representational metaphors that could be manipulated by a real-time data stream, either internally or from external sensors. Here the system creation and development are discussed, up through the design criteria for the final working model that will be used as an example for the efficaciousness of the concept.

1. System Concept Creation

These criteria describe what the system must do: use simple metaphors to create a highly readable visualization containing four or more data dimensions that respond to changes in real time where individual dimensions or general trends can be picked out at will without a great deal of instruction.

Conceptually, this representational metaphor system matches a number of characteristics of ambient displays. Strong, but simple, correlation-based connections link the data to the visualization (unlike arbitrary mappings as with Chernoff faces or the Nabaztag™), where they can be intuitively understood [19]; they can be understood very quickly even if the specific numbers are unavailable; they present the information only peripherally when not needed. However, where most ambient displays are one-dimensional, this system will be expanded into many.

From the review of existing multidimensional techniques, and statistical graphs in general, a number of criteria stood out as relating best to metaphorical displays. First, the ability to quickly show trends or correlations between data dimensions, a strength of the parallel graph, matches the capability of quickly understanding the meaning of one dimension in ambient displays. Second, isolating one or more dimensions for those specific changes, something parallel graphs can't do, though most others can, relates back to the one-dimensionality of ambient visualizations, where every dimension is isolated. Finally, high readability is paramount, the opposite of just about all of the multidimensional techniques described, which require practice, and occasionally instructions, to make sense of them.

2.   System Function Creation

Once access was gained to the historical electrical data from the Langford building complex, the idea of representing information about the building developed. One metaphor that showed a great deal of promise was a landscape. Aside from the humor inherent in using a virtual representation of the natural environment as a metaphor for the weather conditions of the built environment, a landscape is rife with discrete elements that can be altered almost independently without changing the nature of the thing itself.

Developing the landscape metaphor led to creating a list of everything possible that could be measured inside a building (electricity, temperature, light, noise, pressure, etc.) and creating a list of potential linkages between those with aspects that could be visually represented inside the metaphor (wind, cloud cover, tree shape, grass color, etc.). It was important when creating these links that there be some correlation between the data and the display, in order to encourage an intuitive comprehension of the image as a whole. As opposed to Chernoff faces, where data dimensions are essentially assigned at random to available facial dimensions, the intent here was to make each dimension relate directly and conceptually to an aspect of the display.

After the master list was complete, a table was created to describe all the overlaps between the visual dimensions – where changes in one dimension must cause changes elsewhere as well to keep the landscape metaphor intact. The most obvious of these was rain, describing the humidity, responding to changes of the wind, but other cases needed to be considered as well – generally color changes, though some interactions needed shape, light, or shadow changes. Once the mappings and interactions were complete, it was time to build them into a working example.

3.   Prototype Creation

In order to show the effectiveness of the metaphorical system created here, a proof-of-concept must be built according to the previous principles. To ensure that it accurately shows the multidimensional capabilities that were designed into the process, as opposed to merely recreating all of the one-dimensional examples discussed already, the prototype was given four dimensions of data to display, with room remaining for more.

The four dimensions to be used were chosen from the master list created earlier based on two criteria: ease of measurement, and relationship to human comfort. Temperature and humidity are simple enough to measure that watches can be purchased with those sensors built in, making those two ideal candidates. Combined with their importance to comfort, they fit both standards and are most applicable to the metaphor. Due to the VAV system for airflow in Langford, measuring the ventilation is as easy as recording temperature again, making wind the third choice. Finally, electricity was selected for its importance in designating the amount of use the building was experiencing at that particular time, to provide context for the rest of the measurements.

As a functional example of the system, changes must be propagated in real time even if the actual measurements ended up not being possible. Metaphorical coherence was another primary consideration, and finally the composition and visual appearance needed to be strong and pleasant.

CHAPTER IV

DESIGN AND IMPLEMENTATION

This chapter describes the process that eventually took this idea of real-time ambient metaphorical visualization into a functional system and finally built it into a working prototype. Here we develop a metaphor-based visualization of four environmental characteristics of Langford: temperature, humidity, ventilation, and electricity consumption.

The following sections will explain the process of the four metaphorical representations of environmental data, the techniques used to manipulate them with minimal computing overhead, the software used, and finally will relate details on how all the disparate pieces were fit together.

1.   Data Ranges

In order for the simulation to have any sort of reference point, each of the four data dimensions must have an established range. Humidity is simple in this respect, as it's a percentage, and can only run from 0-100%. Electrical usage, temperature, and ventilation level all essentially have no upper limit, however, which introduces a problem. Solving it means making a decision on what levels are important: is it important to know when electricity is at zero, or just when it's as low as it generally goes during a day? Is it important to know when water freezes, or just when it's chilly inside?

These questions were answered in two ways in the construction of this project: first by settling on an intent to demonstrate normal conditions; second by building an interface to allow the user to change the ranges if other information is desired. By default, the system brackets electricity with average minimum and maximum levels, temperature with what feels cold and what feels hot, ventilation with the low state of the VAV system and the average high temperature of the county, and humidity with its entire range.

Regardless of what the actual physical information is, once in the system the values are scaled linearly into a range from 0-100 encompassing the established thresholds. Non-linear scaling could be used in the future to create different visual effects. Whether 0 corresponds to frozen water or 60 degrees it will have the same effect in the simulation. The 0-100 range was used because its step sizes were small enough to appear continuous to the user, while allowing a sufficient range of data display. How wide a spread of real-world data can be described by each of those discrete steps is a function of the spread between the minimum and maximum thresholds.

2.   Visual Dimension Construction

To design the ranges of the metaphors, the possible data values were initially limited to three steps: low, medium, and high. Each of these conditions for each dimension was sketched out several times, until the changes in them were fully developed. Temperature would alter the sky from a dreary overcast to a bright blue summer, while changing the ground from a frozen tundra to baked soil; humidity would dry the grass to straw or saturate it with green, while also triggering rainfall as it got wetter; ventilation would make the grass, clouds and rain blow; electricity would simply increase the amount of lightning throughout the landscape.

Once the stages of the simulation were built in still images, it was necessary to figure out how to expand the stages into animations that could move smoothly from one state to another. The stages that only involve color change are not an issue: as the incoming data changes from state to state, the values used for color multiplication will change just as gradually, producing its own smooth transition. The lightning also posed no problem, due to the abrupt nature of the bolts themselves; they could freely be built into a video clip that could be jumped around almost with impunity from visual artifacts. However, creating smooth transitions between grass blowing and varying levels of rain was an issue. Multiple video files, one for every increment of change, could work, but has enormous overhead in terms of pre-production work and file storage space, as well as being cumbersome to program.

Finally long videos were created for both the rain and grass with progressively increasing quantities of rain and wind. When the system selected appropriate sections of these videos, it was important that the amounts of rain or wind appeared constant. By using a playback window small enough to make the amount of rain or wind appear constant, and balancing the size of the window with the overall change in the video, from minimum to maximum, the problem is solved. Sliding along the video would allow small changes to happen smoothly, but doesn't leave room to change two aspects at once: what about wind blowing the rain? Again multiple video files offer a solution rife with problems. However, if the rain video were created larger than the rest, it would be a simple matter to rotate the entire video clip independently of the playing time to match the grass, a visually effective solution. It's also cost-effective in computer time, particularly when compared with procedurally creating or warping the rain.

One option was to build the videos with discrete steps for each possible state of the system. This turned out to be impractical for several reasons: primarily, it would not allow transitions from one state to another without visual discontinuities, as abrupt changes would be built into the videos. More practically, the videos, metaphor designs, and system construction were all being developed simultaneously. With a continuous video, changes can be made in all areas of development without requiring the complete retooling of the other components – the size of the playback window, the number of values the system distinguishes between, the length of the video, or the total amount of change in the video, for example. While a video containing discrete changes would be physically more accurate, that technique requires all the other pieces to be fully developed first, limits transitional options, and its accuracy isn't even visible to the user. By coordinating the playback window size with the amount of total change throughout the video, the slight changes in the system are invisible.

3.   Interactions Between Visual Dimensions

The previously discussed interactions between all of the data dimensions create a four-dimensional matrix of possible states for the simulation to be in (Figure 5). In order for the simulation to operate in real time, the methods used to solve the interactions (and the changes in state of the metaphors themselves) must be as low-impact on the computer as

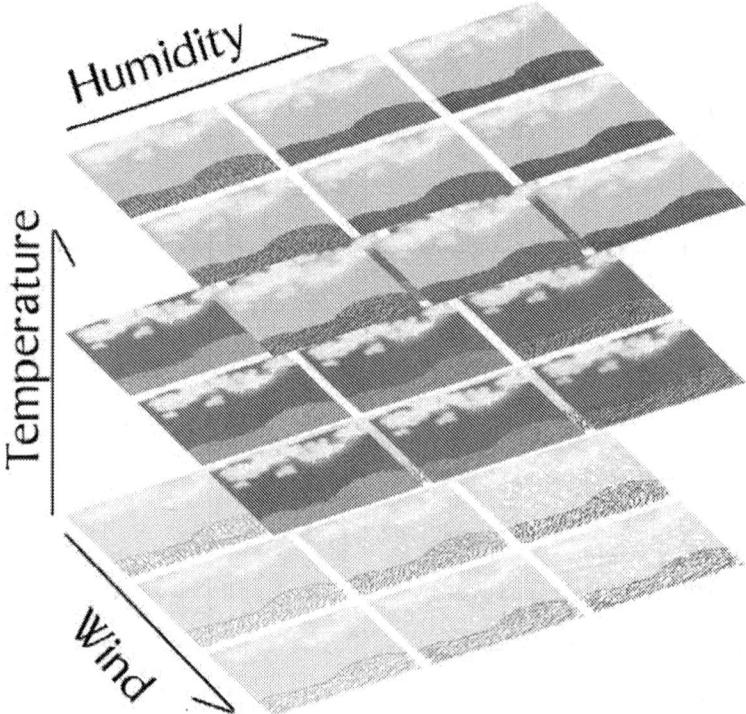

Figure 5. Visualization of the 3D matrix of the possible states of the simulation when electricity is held constant (in order to reduce it from four dimensions).

possible. The less processing power each dimension requires, the slower the computer the system can run on while still reacting in real time, and the greater the number of metaphors that can be used.

Possibly the easiest operation for a computer to perform is multiplication, which is the reason this prototype was built through compositing. With this in mind, one simple technique was settled on for the majority of the dimensional interactions: color multiplication. By individually controlling the colors of each metaphor through multiplication, most of the metaphorical overlaps can be accounted for without a great deal of computing overhead. Some other interactions can be achieved simply by altering the playing speed of the metaphor's layer. Finally, the rotation and transposition of video footage is a relatively easy task, and can be done independently of playback. As it turns out, those three methods were enough to solve all of the overlap in this prototype.

4. Data Sources

Three types of data sources are used in the visualization prototype. Live sources, with the help of small sensors, provide real-time information about conditions. Artificial sources use old or estimated data to provide a simulation of conditions in cases where real sensors can't be used, such as with the electrical data. Finally, the user acts as a source that can always override the others with manual control.

Simulated sensors are treated as though they were live sources. They are queried at the same intervals and interpreted the same. An isolated component of the system independent of the other operations or direct measurements, numbers are generated by interpolating continuous data from historical hourly measurements.

The manual controls directly alter the state of the system in each dimension from low to high, bypassing the threshold modification. Though not truly a data source in its own right, the manual controls still provide facilities for testing and demonstration.

5.   Visualization Software

The creation of the prototype required several software programs: Adobe AfterEffects 5.5™, Alias's Maya 6™, and MetaCreation's Bryce 4™. AfterEffects™, a powerful video processing program, was used to alter the resources coming from the other programs to put them all in the same format, and also created the lightning and rain through some of its filtering capabilities. Maya™ has extensive rendering and dynamics capabilities, as well as the industry standard 3D interface; it was used for landscape and camera placement, applying texture to the ground, and the creation and manipulation of the grass. Bryce™ is, in a manner of speaking, an amateur's Maya™: cheaper and less adaptable, but with its own set of strengths. In particular, because it's meant for users without professional skills, it has powerful tools to simplify the creation of complex features like terrain and clouds. Because a convincing ground and sky were important to the project, that is precisely what Bryce™ was used for.

The actual assembly of the prototype was done within one more software suite: Max/MSP™. Max/MSP™ is an extremely flexible visual programming environment designed for signal creation, acquisition, and processing – originally designed for MIDI music production, it proved useful for all types of signals and has since been expanded to include powerful video processing tools with the Jitter™ plug-in. Because of its adaptability and range of capabilities, it's been used as the software control for many artistic installations.

6.   Visualization with Real-Time Video

The visualization is built as a set of layers of video, each of which corresponds to a dimension of data and an associated metaphor, which can be manipulated independently before being composited.   Each metaphor dimension has its own particular needs, requiring individualized attention to the construction and manipulation for each of the layers.

6.1. Internal Changes in the Video Layers

Electricity is the simplest of the dimensions, as it affects nothing but itself. The amount of electricity fed to the system corresponds only to the frequency of lightning strikes. The number of lightning strikes within the video clips the system plays numerically demonstrate the quantity of electricity being used, while visually the perceived magnitude of the lightning strikes (how close they are to the viewer, and whether they strike the ground or not) relates to a rougher estimate of the power consumption of Langford.

The electrical metaphor consists of a single video file composed of increasingly frequent lightning strikes created in AfterEffects™, thirty seconds long and sectioned in two different ways: visually and numerically (Figure 6). Both types of segmentation increment along with the time of the video file, allowing the Max patch to simply slide a five-second playing window up or down the file, relative to the current value of electricity use. Because of the abrupt nature of lightning, blips that occur from inexact loop points aren't noticeable.

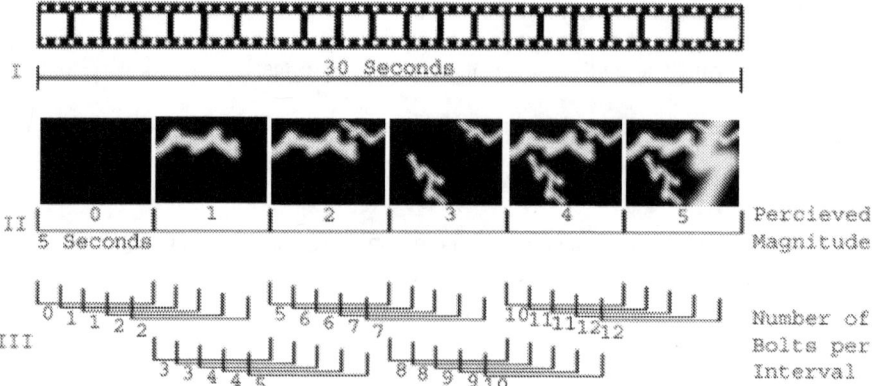

Figure 6. The lightning video clip. The five-second playback window used for the
lightning layer contains two types of information: the perceived magnitude of
the lightning strikes, as well as the actual number of lightning bolts contained
in the playback window.

The video contains six visually distinct segments, dividing the clip every five
seconds and influencing the perceived magnitude of the lightning (Figure 6, II). They
are built with the idea that distant lightning is less threatening than close lightning, and
lightning that hits the ground is more severe than that which doesn't. The first section,
from seconds 1-5, is the lowest level of electrical usage, so contains no lightning at all.
Seconds 6-10 contain only sheet lightning – cloud-to-cloud lightning with no ground
strikes – in the middle distance. Seconds 11-15 have sheet lightning in the middle and
far distances, frequently overlapping. In the fourth segment, from 16-20, the middle-
distance sheet lightning stops but ground strikes begin along the horizon. Seconds 21-25
contain middle-distance sheet lightning again, as well as the distant lightning. Finally,
the highest amount of electrical usage relates to the sixth segment of the video, which

adds a set of ground strikes very close to the camera, allowing up to four simultaneous bolts of lightning. As later segments of video are used, the perceived magnitude of the lightning becomes increasingly intense, allowing easy estimation of trends.

The visual breakup of the video coincides with a numerical sectioning based on the quantity of lightning (Figure 6, III). Beginning with none, every other five-second piece of the lightning video contains one more lightning strike than before. Along with the visual key, this simultaneously sections the lightning into both six and thirteen segments. In this way a high degree of numerical accuracy is possible, without disrupting the metaphor.

Ventilation is represented as wind, which is invisible on its own. Grass was added as an ever-present medium to react to the wind. (Other aspects of the simulation have to react to it as well, or the metaphor would collapse, so when it's present, rain is affected as well, as are the clouds.) A thirty-second video clip of grass blowing on a black background was made using the dynamics in Maya, with the wind strength increasing from nothing to an amount that has the blades bent almost completely over. A two-second window of playback time is used, to get the minimum amount of wind change with the time while not getting too repetitive. Because the window can't be guaranteed to loop smoothly at every point along the ventilation scale and the motion is essentially the same backwards and forwards, the video "palindromes" – loops forward and backward.

The grass isn't well suited to the sort of exactness that the lightning achieved, as it isn't nearly so discrete. However, the video clip itself can be broken up into similar segments and the resulting angles of the grass blades can be used as a key that the user can consult and estimate with practice.

Temperature is represented directly by two features of the landscape: the color of the ground beneath the grass and the color of the sky. As with the grass, direct numerical extraction is difficult. Also like the grass, artificial gradations and color keys are simple to make.

The ground color is altered by a simple color multiplication. Three key colors were chosen for the color to be multiplied: a pale, barren color for the coldest point, a neutral brown for the mid-point, and a deep terracotta for the hottest level. Linear interpolation is applied between these, creating a gradient between them proportional to the registered temperature.

As the registered temperature increases from the low threshold to the high, the sky fades from a dreary overcast gray to a clear blue to a deep blue with tinges of red. This is accomplished with a 100-frame video clip of the sky fading between those colors created in Maya™, where the temperature reading dictates a single frame to be displayed. This technique, as opposed to the ground's flat color multiplication, allows the sky to have texture, haze, and horizon effects without being computationally intensive.

The humidity fed into the system determines the amount of rainfall present, as well as how dry the grass is. Below the median humidity set in the system no rain falls and the grass is the only indicator. Above that median threshold rain appears, in a quantity proportionate to the humidity level.

The rain consists of a thirty-second video file created with AfterEffects™ that, like the grass and lightning, begins with no rain and incrementally adds more. The rain-production algorithm used by the filter created very obvious changes in the amount of rain at very low levels, so an exaggerated ease-in curve was used to force more even rain-production. The speed of the rain works just like the lightning video to mask the loop points of the video, and keeps the two-second playback increments from becoming repetitive. Such a short video segment was used because it works visually, to ensure a constant level of rainfall for any given humidity level.

6.2. Compositing Video Streams

Within the Max/MSP™ patch there is still need for a user interface. Controls, aside from start and stop buttons, are kept in the background, but are available if needed. The simplest of these are the manual controls, which simply override the sensors to allow manual data input. More advanced settings allow the user to set thresholds for each of the dimensions in order to focus on more specific areas of interest: some amount of electricity will always be used in a building, so it's useful to set the minimum range of the electrical dimension near the baseline of the building to get the maximum information available.

Behind the interface, the Max™ patch performs some deceptively simple operations: choosing what points of each video layer to play, altering their colors, and stacking them on top of each other.

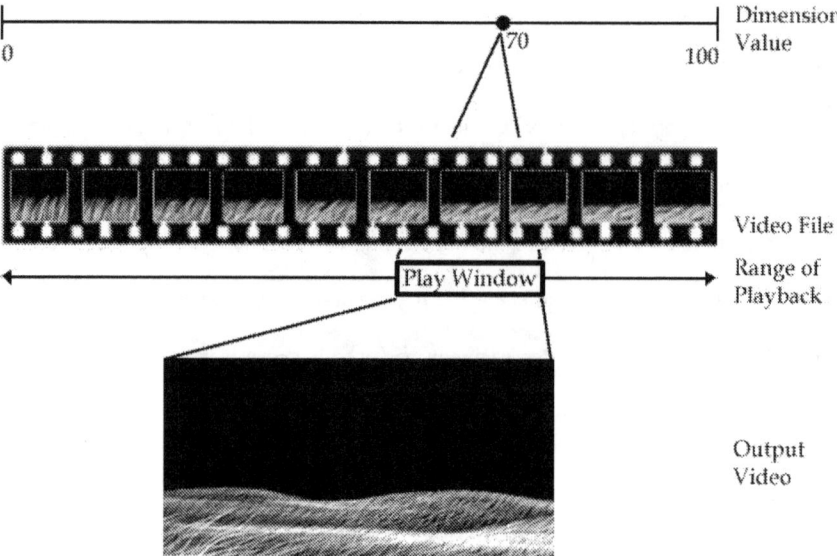

Figure 7. The value of the data dimension determines linearly where the playback
window will be placed along the video file. This window sets the loop points,
looping style (normal or palindrome) and outputs the video layer. This process
is repeated independently for each visualization dimension.

Because the patch has been programmed with the absolute length of each video

file, and knows how much of each it should play at a time, determining which section of

each video the patch displays is calculated from a linear interpolation of the value of the

dimension across the range of possible starting frames (the length of the video less the

size of the playback window) (Figure 7). Because the component videos have steady,

continuous change rather than discrete stages, any one of the factors (length of the video,

length of the playback window, and amount of total change in the video) can be changed

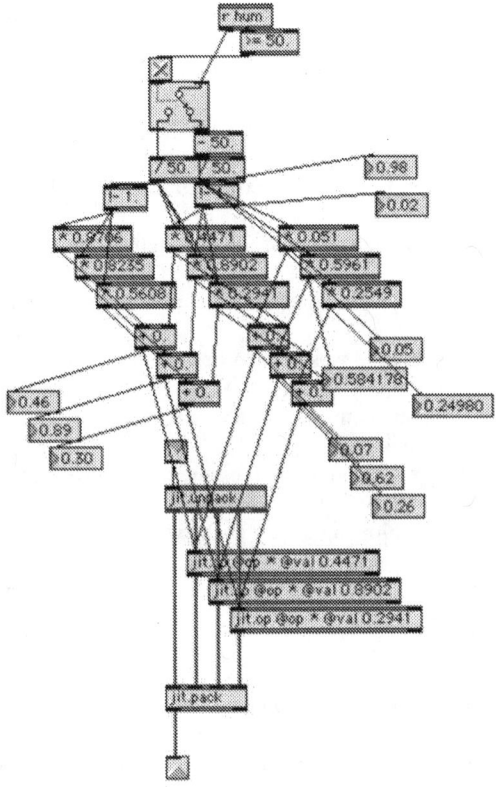

Figure 8. The color interpolation and multiplication that determines the final color of the grass.

without disrupting the rest. As the design and implementation of both the visual and programming aspects were happening concurrently, this proved an advantage.

Changing the color of each layer first requires deciding upon three key colors to use: the minimum, middle, and maximum values. The patch interpolates linearly between them according to the value of the dimension the color represents to create the red, green and blue values that will actually be multiplied. This new color is then multiplied across the video frame (Figure 8). Black, being values of zero in all three channels, remains unchanged, while white, being values of one in all three channels, becomes the new color exactly. All the shades of gray in between become darker versions of the new color, keeping the texture and shading of the frame intact.

The heart of the visualization of the prototype system is compositing. Six layers of prerendered video are layered together in real time through simple multiplication computations. This technique was chosen because it allows a great deal of realism and complex behavior without requiring a lot of computing power. The sky, technically a complex gradient change with additional texture can be processed as easily as displaying an image, while the grass blowing gently in the wind – a process which takes upwards of twelve hours to render – can be adaptively played on less powerful computers without any trouble.

While real-time compositing of prerendered elements allows for a lot of normally intensive visualization, it is also limited by a lack of flexibility. Very few of the pieces of the simulation can be changed without requiring changes to all the rest. The camera can't move and the ground can't change shape or every other piece is disrupted; any application that requires a freely moving viewpoint would be poorly suited to this technique. In the case of this prototype, which is meant to be a freestanding installation with minimal controls and ambient information, these restrictions aren't an issue.

Compositing is accomplished through simple multiplication equations through an alpha mask or channel. The alpha channel ranges from 0-1 with zero meaning perfectly transparent and one completely opaque. If you stack two images with alpha channels and want to merge them, then for the color of each pixel the algorithm is simply:

$$C_3 = \alpha_1 * C_1 + (1 - \alpha_1) * C_2$$

α_1 represents the alpha value of the top image, and C_1 encapsulates all three channels, red green and blue of the top image, C_2 the composite color of the bottom image, and C_3 the resulting color values. When implemented, the equation is repeated three times on each pixel of the image, once for each color channel.

The new alpha value, α_3, of that combined pixel is determined by the equation

$$\alpha_3 = \alpha_1 + (1 - \alpha_1) * \alpha_2$$

All of the video file layers have been created against a completely black background, and in black and white where it was possible. This way an alpha mask can be constructed by duplicating one of the other color channels (before the colors are altered for display) such that the black areas are transparent, and the white, where the actual subject matter is, are opaque. Because white is the result of all the color channels being equal, any of the channels can be selected with the same result. Alternatively, alpha masks could be carried in separate video files dedicated to that task, but the computing overhead of running two video files compared to using one file twice outweighs the level of control prerendered alpha channels could give. Additionally, creating an alpha mask on the fly through channel duplication has the advantage of

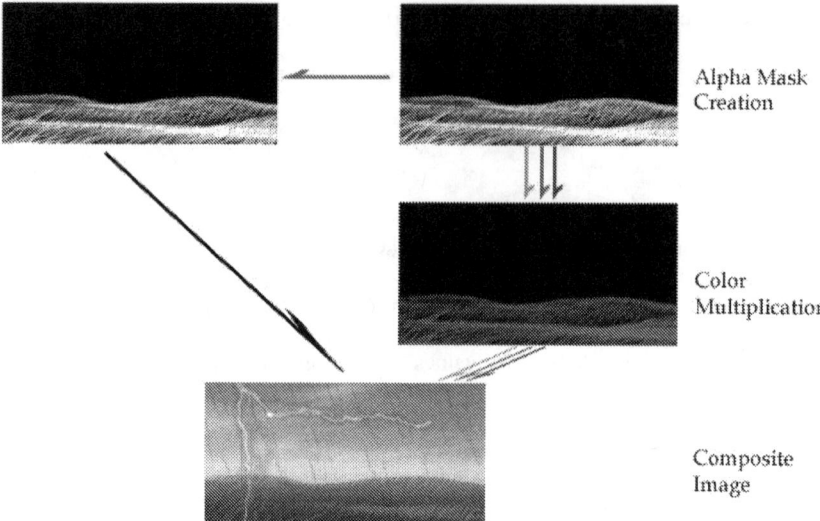

Alpha Mask
Creation

Color
Multiplication

Composite
Image

Figure 9. The alpha mask is created from the duplication of any of the color channels. Because the image is black-and-white, all channels are equal, and red was arbitrarily chosen. The Max patch performs a color multiplication on the red, green, and blue channels, then uses the newly colored image combined with the alpha mask of the current frame to create the final composite image.

always being correct: the current frame and the current alpha mask will always match, without requiring careful attention to playback speeds and times (Figure 9).

Each of the layers of video plays at least one part in the final composition. The sky is, naturally, the very bottom layer, and fills up the rest of the frame remaining from the other video pieces. It's directly controlled by the temperature, and built in a way that it fades to a paler color at the same place as the horizon line, creating a haze effect.

The landscape is the second layer, to take into account that all the rest of the pieces occur above ground. Aside from being the primary compositional element of the final assembly and providing context for the grass and lightning, it also changes color based on the temperature, via color multiplication. The base, rendered image of the land is grayscale, with speckled noise to texture it and a distance-based fade to darken it in the foreground, increasing the depth of the image to avoid the image becoming flat or cartoony. The multiplication operation ensures that the dark foreground will be a deeper color, while the white background will be pale and appear hazy, with the entire process still resulting in the selected hue.

A number of different terrain styles were made before the final version was settled on (Figure 10). The first were arbitrary constructions of land shapes created solely to show depth. Following those, attempts were made to duplicate the terrains of classical landscape paintings, generally resulting in bold, rugged hills and valleys. Finally, it was decided that because this installation was built to represent a building in eastern Texas, and because the shape of the ground wasn't an active data representation itself, the virtual land should be a reflection of the real land surrounding Langford: gently rolling plains. This final landscape was built in Bryce™, where special tools exist

Figure 10. Some of the first attempts at the landscape, determined to be too dramatic and inappropriate for a representation of Langford.

just for landscape creation, and imported into Maya™, where far more powerful texturing and rendering tools could be used.

Like the hills, the grass was prerendered white against a flat black background and invisible hills. By copying any of the color channels as an alpha mask (equivalent to using luminosity because each of the color channels are identical), everything black becomes transparent, while everything gray is translucent and the white is completely opaque. The humidity measurement controls the color of the grass – from a pale, dry brown to a deep saturated green – using the same color multiplication as the hills. It naturally fades into the distance because of the rendering algorithms, the grayer edges add texture and fluctuations to the green, and it works very cheaply and effectively.

The lightning video is composed in the same way as the grass and hills: primarily black or transparent, save the lightning itself. There is one difference: this is the first case where the subject matter has color of its own (pale yellow). Because the color of the lightning was not going to change, it was created already tinted to limit processing time. It's also surrounded by a mostly-transparent glow, which must be taken into account when creating the alpha mask. In the RGB color scheme, yellow is created as a combination of green and red, meaning either of those channels would be a good choice for alpha mask creation. However, the core of the lightning is white, or equally bright across all three colors, and it was important for the yellow glow to be subtle. To this end, blue was chosen to become the alpha channel.

Above the lightning are the clouds. This layer was created with two programs: Bryce 4™, which includes very simple but powerful controls to simulate very believable cloud cover; and AfterEffects™, which was used to alter the cloud video such that it loops smoothly. This video clip plays constantly as the simulation runs to give the sky texture and keep it from being too static. Though the cloud cover doesn't actually correspond to any data dimension (in this iteration of the work – future work in this area may take advantage of clouds), in keeping with the metaphor they do still respond to the wind strength from the ventilation; as the wind increases, so does the play speed of the cloud video.

Contrary to the previous examples, Bryce™ did not allow black backgrounds while simulating atmospheric effects like clouds, so the rendered video in this case includes a blue background. It's important to be sure none of the sky from the video interferes with the sky layer for the temperature, so all of the blue must be removed. While the sky is primarily blue, it is also partially cyan – a combination of blue and green – which means the red channel, duplicated as the alpha mask, most completely removes all of the sky Bryce™ introduced.

One more feature was added to the video clip of the clouds to improve the composition of the whole work: a gradient was applied to the lower half of the video file, fading from black at the base to a thick white bar where the horizon of the ground plane appears once the files are layered together. Because of the multiplication used while compositing, this creates the illusion of atmospheric perspective – the more distant hills are viewed through a haze, giving a much better illusion of depth then previously achieved.

The final layer of video is the rain, which returns to the format of the grass: white and gray raindrops on a transparent black background. The drops change in color to correspond with the temperature, moving from a freezing rain to warm drops. This video file is also larger than the rest, so it can rotate in coordination with the wind without the edges of the video appearing within the composition.

CHAPTER V

RESULTS

The final work operates and updates in real time, and effectively composites all of the video pieces together, colorizes everything based on the data fed into the system (Figure 11).

Of the six compositional layers, most have more than one degree of change (though not all are implemented in the prototype, the clouds and lightning could have color changes applied to them very simply if appropriate dimensions were found for those representations). Only the sky and ground layers, which simply change color, would require major changes to themselves and elements around them to add other degrees of freedom. The rain layer actually has three degrees of change: the quantity of rain, the color of the drops, and the angle at which they blow. These allow all the changes necessary for the four-dimensional metaphor system that was built.

The development of the conceptual system showed a great deal of flexibility. The dimensions finally chosen combined together well to create an effective and adaptable result.

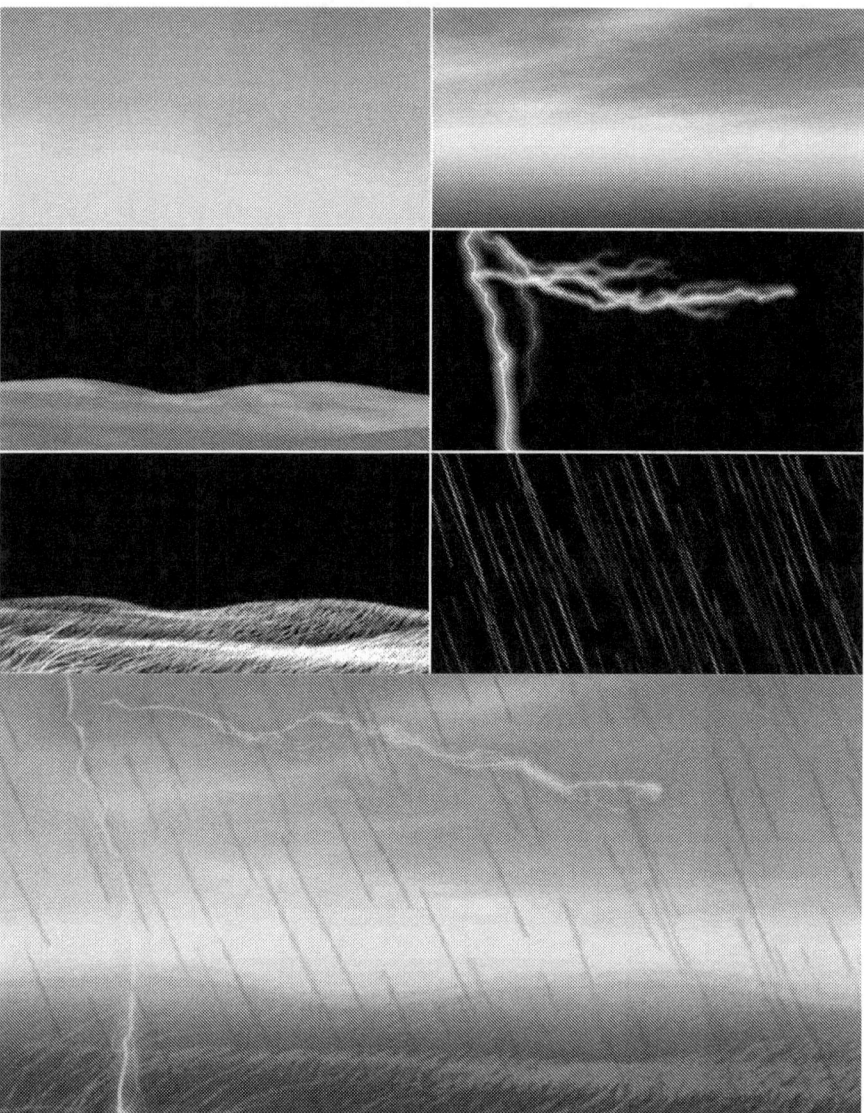

Figure 11: The six, generally black-and-white images, combined and colorized into the final composition.

CHAPTER VI

CONCLUSIONS AND FUTURE WORK

The final iteration of the prototype operates smoothly in real time, handling all of the user inputs, data sources, video playback and dimension interactions without trouble on an Apple G5™. It demonstrates how ambient metaphorical displays could be used and understood, and suggests a number of advancements that could be added to it.

Because the prototype was designed from criteria selected from previous work to be highly readable, it is suspected that this work can be read easily and intuitively. However, future work must include perceptual studies to that effect to definitively make such a claim.

A strength of this approach is the ability to show interactions – not just artificial interactions within the metaphors to keep the illusion intact, but also potentially important interactions within the data itself. Because all of the data dimensions are processed simultaneously, important combinations of data can trigger special events when two dimensions individually are within acceptable ranges but the combination of the two requires attention.

As an installation, this prototype would be effective as a public demonstration of conditions in a building, as an ambient meter for environmental systems technicians, or perhaps most interestingly, as a comparison of remote locations by separating the sensors from the visualization itself.

Obvious improvements to this proof-of-concept involve creating different top-level metaphors to fit the rest in – or simply developing a less realistic rendering of the environment to open other metaphor possibilities – and the addition of more dimensions, with more metaphors. Trees or buildings or animals all could correspond to measurements not proposed here; the clouds could become a direct data dimension, or the shape of the ground could relate to something.

Changing either the clouds or the terrain in real-time without a supercomputer would probably require the development of an exotic video format, such as a two-dimensional frame structure, where playing one direction of the frames shows the passage of time, while playing the other shows the changing of shape. With some clever interpolation between the frame dimensions, diagonal directions could be played, changing the rate of change through time by changing the angle, allowing every possibility to be pre-rendered, creating seamless morphing from one state to the next.

The prototype could be expanded to have a number of copies networked together, creating a larger data net and showing how conditions change between different locations. The passage of time could be expanded upon to add more context, either through daylight and nighttime, or by building in changes due to the seasons. Some amount of forecasting could be built in using weather data online to predict changes that might occur later, perhaps displayed as storm brewing on the horizon.

REFERENCES

[1] E. Tufte, *The Visual Display of Quantitative Information.* Cheshire, Connecticut: Graphics Press LLC, 2001.

[2] E. Tufte, *Visual Explanations.* Cheshire, Connecticut: Graphics Press LLC, 1997.

[3] E. Wegman, "Hyperdimensional Data Analysis Using Parallel Coordinates," *Journal of the American Statistical Association*, vol. 85, pp. 664-675, 1990.

[4] S. Fienberg, "Graphical Methods in Statistics," *The American Statistician*, vol. 33, pp. 165-178, 1979.

[5] H. Chernoff, "Using Faces to Represent Points in K-Dimensional Space," *Journal of the American Statistical Association*, vol. 68, pp. 361-368, 1973.

[6] C. Bajaj, G. Xu, J. Warren, "Acoustics Scattering on Arbitrary Manifold Surfaces," *Proceedings of Geometric Modeling and Processing, Theory and Applications*, p. 73, 2002.

[7] C. Wisneski, H. Ishii, A. Dahley, M. Gorbet, S. Brave, et al., "Ambient Displays: Turning Architectural Space into an Interface between People and Digital Information," *Proceedings of the First International Workshop on Cooperative Buildings*, pp. 22-32, 1998.

[8] S. Card, *Readings in Information Visualization: Using Vision to Think.* San Francisco: Morgan Kaufmann Publishers, 1999.

[9] P. O. Fanger, *Thermal Comfort: Analysis and Applications in Environmental Engineering.* New York: McGraw-Hill Book Company, 1970.

[10] H. Wendes, *Variable Air Volume Manual*. Lilburn, Georgia: The Fairmont Press, Inc., 1994.

[11] G. Lakoff, and M. Johnson, *Metaphors We Live By*. Chicago: University of Chicago Press, 1980.

[12] T. Winograd, *Bringing Design to Software*. Boston: Addison-Wesley Professional, 1996.

[13] S. Eick, "Data Visualization Sliders," *Proceedings of the 7th ACM Symposium on User Interface Software and Technology*, pp. 119-120, 1994.

[14] E. Wells, L. G. Tassinary, "Standardized Performance Trajectory as a Measure of Usability," *Proceedings of the 4th Annual Symposium on Human Interaction with Complex Systems*, pp. 226-234, 1998.

[15] A. Galloway, "Carnivore," http://rhizome.org/carnivore/, 2005.

[16] N. Jeremijenko, "Database Politics and Social Simulations," http://tech90s.walkerart.org/nj/transcript/nj_01.html, 1997.

[17] N. Gershenfeld, *When Things Start to Think*. New York: Henry Holt and Company, 1999.

[18] H. Ogawa, N. Ando, and S. Onodera, "SmallConnection: designing of Tangible Communication Media over Networks," *Proceedings of the 13th ACM International Conference on Multimedia*, pp. 1073-4, 2005.

[19] D. Norman, *The Design of Everyday Things*. New York: Basic Books, 1988.

VITA

Eric Brian Aley
752 Espolon Dr.
El Paso, TX 79912
ealey@tamu.edu

Education

M.S. in Visualization Sciences	Texas A&M University, May 2006
B.E.D.	Texas A&M University, May 2003

www.ingramcontent.com/pod-product-compliance
Lightning Source LLC
LaVergne TN
LVHW080105070326
832902LV00014B/2426